YOUR KNOWLEDGE HAS VALUE

- We will publish your bachelor's and
  master's thesis, essays and papers

- Your own eBook and book -
  sold worldwide in all relevant shops

- Earn money with each sale

Upload your text at www.GRIN.com
and publish for free

Nutrition and its effect on Productivity at the Workplace

John Ramotowski

Bibliographic information published by the German National Library:

The German National Library lists this publication in the National Bibliography; detailed bibliographic data are available on the Internet at http://dnb.dnb.de.

ISBN: 9783346864512
This book is also available as an ebook.

John Ramotowski

Nutrition and Productivity

MASTERS THESIS

American College of Healthcare Sciences

Abstract

This document explores the relationship between nutrition and productivity at the workplace and, to a lesser extent, lifestyle. Nutrition is essential for the body to sustain itself and enable its systems to work productively. As important as productivity is, it can be affected by things such as immunity, energy balance, sleep, and more which can be significantly assisted by nutrition. The sections of wellness points described below can affect levels of productivity. The discussion takes place in this document on how diet can affect each point. The topics are Travel and Vacation and how to fight illness before leaving on vacation, Combating stress with nutrition, concentration, and focus on nutrition, anxiety, fighting disease, and the immune system, energy highs and lows, sleep, food at work, and meetings, and emotional eating.

Methods used to gather information were conducted by searching various databases such as PubMed, Google Scholar, Google, and various medical internet sites. Included in the methods were people of the general healthy working population, mainly in the United States, and exclusions were those with terminal illnesses. The results for each point are in the document, and the discussion allows for the best actions to help productivity thrive at the workplace and personally.

Keywords: productivity and nutrition, workplace productivity, emotional eating

Nutrition and its Effect on Productivity

People have specific dietary needs, and diet provides for these needs. Healthy food provides a well-balanced diet for the body. Inadequate nutrition can harm the body in many ways, like a weaker immune system, more access to becoming ill, loss of mental and physical development, and decreased productivity.

Appropriate nutrition is needed to maintain proper wellness and physical endurance and keep illness at bay. People with suitable dietary needs met by consuming whole foods, whole grains, leaner meat, and fish have a stronger immune system to fight disease and fewer diseases caused by a poor diet. Foods include fresh fruit and vegetables, organic meats, and wild fish versus farmed. The Standard American Diet (S.A.D.) contains high amounts of saturated fats, hydrogenated oils, refined carbohydrates, and highly processed foods (Leonard, 2013).

Eating poorly and having a sedentary and stressful lifestyle can create an environment for illness and obesity. Numerous diseases relate to this. These include high blood pressure, diabetes, coronary heart, stroke, sleep apnea, osteoarthritis, and certain cancers (Leonard, 2013). Food choices create an outcome that can lead to being ill or healthy and impacts productivity. Making nutritional choices can also result in more robust well-being and self-esteem, higher skills to cope with stress, better overall health, lower health costs, and enhanced satisfaction and productivity at work.

Poor employee health and related productivity losses cost U.S. employers $576 billion per year, including workers' compensation, disability, and group health program expenses, according to research in 2012 (Ceniceros, 2012). Many things can affect workplace productivity. The following are examined and shown how each can impact a person's health and productivity.

These are:

- Dealing with work travel, vacation

- Stress, Concentration, and Focus

- Anxiety

- Flu season

- Energy highs and lows throughout the day

- Sleep and its effects on productivity

- Fitting in proper nutrition during the day at work

- Meetings and nutrition

- Emotional eating and how these tie into all aspects of the body, mind, and spiritual health

As many as 85% of U.S. employers are interested in services to increase employee productivity, minimize absences and improve the health of their employees (5 Factors, 2016). Some estimates show that 18 to 20 million Americans aged 19-64 are not working due to health reasons (5 Factors, 2016). There were roughly 69 million workers who had missing days due to illness last year, for 407 million days of lost time at work (5 Factors, 2016). At the same time, about 40% of U.S. workers are fatigued, which causes people to perform at less than half capacity for health reasons (5 Factors, 2016). Stress leads to employee disengagement, which also negatively affects productivity.

For U.S. employers, fatigue bears estimated costs of more than $136 billion per year in health-related lost productivity, $101 billion more than for workers without fatigue (5 Factors, 2016). Eighty-four percent of the costs were related to decreased performance at work rather than absences. (5 Factors, 2016).

One of the biggest challenges for employees is presenteeism (being present at work but not performing optimally) (Holwegner, 2013). Like many corporations, time management was considered a systems and technology option to improve. After further research, it is found that health-related factors are more significant issues than investing in business resources or time management training (Holwegner, 2013). The bottom line will be severely affected if employees are unproductive and not at work (Ceniceros, 2012).

Addressing the factors behind lower productivity is beneficial. These factors can be about nutrition, mindset, personal problems, attitude, and illness and how diet and lifestyle tie into all these. This paper will discuss how making healthy diet choices can positively affect productivity.

Methods

This document's purpose is to examine how various lifestyle factors could impact productivity in the workplace. Searches for data are from databases such as PubMed, Google Scholar, Google, and ProQuest. Peer-reviewed and full articles are the main contributors. A review of generally employed populations makes up the data for this study. The scope of research will be conducted using current or past written research and previous client experience. Peer-reviewed articles will be significant, and the presentation will occur using PowerPoint. Keywords include and are not limited to: "Nutrition and productivity," "Nutrition and wellness," "Lifestyle and Productivity," "Stress and productivity," "Sleep and productivity," "Obesity and productivity," and "Emotional eating and productivity."

This project will gather data that will discuss how nutrition, lifestyle, and productivity at work relate. The main component of the data is derived from the American working population, and exclusions are those who are terminally ill. This type of research is Quantitative because the study will seek to explain the causes of the concern at hand through objective measurement and quantitative analysis or statistics (Jefferies, 1999). Because qualitative research is more concerned with understanding what is happening as viewed by the participants, it is different from quantitative (Jefferies, 1999). This research is also qualitative from the emotional eating perspective. It ties into the notion of the determinants of food and lifestyle that individuals choose: biological, economic, physical, social, psychological, and attitudes.

Results

Travel and Vacation - Combating illness before leaving on vacation

The body continuously produces stress hormones, such as adrenaline, to help fight infection. Once the body relaxes, the levels decrease and make the body susceptible to illness (Jones, 2014). Evidence shows that eating well before vacation, catching up on sleep, and taking specific vitamins or probiotic supplements could benefit the immune system (Jones, 2014).

Hydration. Drinking at least one liter of water every five hours is a recommendation from researchers from St. John's University in Taipei (Brown, 2014). Dehydration on flights occurs because people do not drink enough water, and the cabin microclimate increases the dehydration rate (Brown, 2014).

Melatonin. The medical consensus on the effects of melatonin, the hormone that controls sleeping and waking cycles on jet lag, is mixed (Brown, 2014). Studies have found that it does help overcome jet lag. Various studies have monitored the effects of melatonin on long-haul flight passengers and found that those who took the hormone improved sooner than the placebo group (Brown, 2014).

Immunity and travel. Building up the immune system before a long flight is significant. Sugar suppresses the immune system, whereas fruit and vegetables are immune-boosting (Brown, 2014). Intensely colored fruit and vegetables, such as dark green, red, blue, and black, and not overcooked, are beneficial. Also, airplane food is always very salty, which encourages water retention in the body. Potassium-rich vegetables will combat this, as well as vegetables such as celery, carrots, cucumbers, and some organic almonds, as they contain magnesium (Brown, 2014).   Astaxanthin supplementing is beneficial for its antioxidant properties (Bowman, 2014). It can improve blood flow, which is essential when sitting on a plane, and it can lower oxidative stress, which benefits the immune system (Bowman, 2014). Studies show it has the highest antioxidant activity against free radicals (Bowman, 2014).

Stress and travel. Stress plays a significant role in air travel. Besides a healthy diet, restorative sleep, regular exercise, and adding essential nutritional supplements to the routine are helpful (Kroner, 2011). One mineral that helps to combat stress is magnesium. It is one of the first nutrients to be depleted during stress (Kroner, 2011). The adrenal glands depend on magnesium, as do over 300 different enzyme reactions in the body (Kroner, 2011). Taking 100mg of magnesium taurate in the morning of the flight is recommended, and another 100mg just before the flight (Kroner, 2011).

Stress

Stress has many effects on the body, leading to poor workplace productivity. Repeated acute anxiety and persistent chronic stress can contribute to inflammation in the circulatory system, specifically in the coronary arteries, which ties stress to a heart attack (Stress effects, 2016).

Stress, in biological terms, refers to the aftereffects of a person failing to respond adequately to an event that has occurred in their life, whether physical or emotional (Ritchie, n.d.).   Stress happens in three stages. The first is an initial state of alarm which produces an adrenaline rush in the person's body (Ritchie, n.d.). The second stage is a short-term resistance mechanism that the body sets up to cope with the problem, and the final stage is a state of exhaustion in the body which can lead to adrenal fatigue (Ritchie, n.d.).

With the sudden onset of stress, the muscles tense up all at once and then release their tension when the pressure passes (Stress effects, 2016). Chronic stress causes the muscles in the body to be constantly guarded. When muscles are tense for extended periods, this may trigger other body reactions and even promote related stress disorders (Stress effects, 2016). A tension-type headache and migraine headache refer to muscle tension in the area of the shoulders, neck, and head which can cause a very unproductive workday (Stress effects, 2016).

One of the main issues with stress is that it can create unhealthy eating habits, which applies to people who are always on the go and lead busy lifestyles (Ritchie, n.d.). People here often endure significant stress and have no time to fit in balanced nutrition around their busy schedules (Ritchie, n.d.). Stress makes the body crave foods high in fats and sugars, and this concern with eating will inflict a greater stress on the body and additional problems that threaten physical and mental health (Ritchie, n.d.).

A common reaction when a person becomes stressed is a sudden urge to eat food. These foods consumed in this situation are mainly convenience foods considered a quick fix to nullify stress and worsen the problem (Ritchie, n.d.). The most crucial issue is the harm that stress can inflict from inadequate nutrition. If there are bad practices in food management while under stress, there is a high risk of severely damaging the body (Ritchie, n.d.).

When someone is stressed and does not eat the right amount of food or nutrients, they will encounter inconsistencies in their blood sugars (Ritchie, n.d.). These differences lead to the person not behaving as usual, which may result in tiredness, lacking concentration, and mood swings (Ritchie, n.d.). If stress is not dealt with properly in the short term, the body will suffer in the long term with blood sugar problems that can lead to diabetes (Ritchie, n.d.).

The link between stress and nutrition is vital. A balanced diet plays an important role when we are under pressure as well balanced nutrition plan will boost resistance against the effects of stress on the body (Ritchie, n.d.). When under stress, the nutrients can be depleted much quicker than usual, and the body begins to degenerate. Thus, it is essential to top up vital nutrients to cope with stress regularly. (Ritchie, n.d.).

Nutrients required include minerals, vitamins, proteins, healthy fatty acids, and energy from foods containing carbohydrates, protein, and fats. Consuming the nutrients typically between 40 and 60 per day is essential to a healthy, well-protected body (Ritchie, n.d.).

Avoiding some foods can be beneficial. Foods containing large amounts of sugar contain no vital nutrients that are required (Ritchie, n.d.). Sugar also gives the person a massive burst of energy for a short period, and when this high runs out, the person can suffer an extended low period (Ritchie, n.d.).

B vitamins are essential for coping with stress as they are used in building up the metabolism, and substances like alcohol and caffeine will drain these resources and affect the brain's functionality (Ritchie, n.d.). Caffeine can be responsible for inducing the first stage of stress, the alarm Stage, and when consumed under stress, the body uses reserve B vitamins, so there are no resources for coping with the problem (Ritchie, n.d.). Caffeine is also responsible for making people hyperactive and nervous, and the person's sleeping pattern is affected significantly (Ritchie, n.d.).

Shift workers are at higher risk of various metabolic disorders and diseases such as obesity and diabetes (Lowden, Moreno, Holmbäck, Lennernäs, & Tucker, 2010). There can be links between diet quality and irregular eating timing. However, other factors that affect

metabolism are likely to play a part, including psychosocial stress, disrupted circadian rhythms,

sleep debt, physical inactivity, and insufficient time for rest and revitalization (Lowden et al.,

2010).

Recognizing nutrients that are used up is vital, and replenishment is critical. The

following are the primary nutrients the body will use (Ritchie, n.d.).

Vitamin B group: These help the body cope with stress, build metabolism, and control the

whole nervous system.

Proteins: Assist in growth and tissue repair.

Vitamin A: Essential for normal vision.

Vitamin C: Protection of the immune system and lowers the amount of cortisol in the

body.

Magnesium: Needed for various tasks such as muscle relaxation, fatty acid formation,

making new cells, and heartbeat regulation.

Foods with poor nutritional value can lead to stress on the body. Their nature is to

stimulate the body, even when it does not have the resources available for additional stimulation,

and this is why these foods and drinks directly cause stress (Ritchie, n.d.). A short burst of

energy may occur after consuming foods that stimulate and, in the long term, can damage the

body (Ritchie, n.d.).

The ingredients listed below may stress the body (Ritchie, n.d.).

Caffeine. Beverages such as coffee, tea, and carbonated products contain caffeine.

Caffeine causes adrenaline to release into the body, which causes stress. If caffeine is taken in small doses periodically, it can be effective as it alerts the body. Too much caffeine can cause high stress, high blood pressure, and nervousness.

Sugar. Sugar contains no vital nutrients, and too much sugar can force the body to use reserved resources, resulting in short-term problems such as lack of concentration and fatigue. Sugar also causes the pancreas to be overloaded, which may lead to issues such as diabetes.

Fats. In moderation, fats are essential to the body. High consumption of foods that contain large volumes of saturated fats can lead to obesity, heart conditions, and cancer.

Alcohol. Many people may drink alcohol under stress, which has the reverse effect on the body. Alcohol will make the body release significant amounts of adrenaline that result in nervousness, lack of sleep, anxiety, and skin irritation. Alcohol will also increase the number of fatty deposits around the heart when consumed under stress.

Smoking. Cigarettes contain nicotine, which people take as a sedative. Smoking would appear to remove the presence of stress in the short term, but the damage done will reappear in the long term (Ritchie, n.d.). Perceived short-term smoking benefits are affected by several long-term problems. Smoking is responsible for some cancers, tension, breathing illnesses, and heart disease.

A prolonged healthy, nutritional eating period will prepare the body for future stressful events (Ritchie, n.d.). The next time this occurs, the body will be more equipped to handle the situation. When under stress, it is important to consume all the necessary nutrients to cope and function effectively (Ritchie, n.d.). The following nutrients are essential (Ritchie, n.d.).

- B Vitamins: There are many B vitamin types, each as important as its counterpart. B vitamins are in foods such as seaweed and raw foods.
- Proteins and Iron: Meats, eggs, seeds, and nuts.

- Vitamin A: Cheese, eggs, fish with oil, and milk.

- Vitamin C: Fruits such as apples, bananas, and oranges

- Magnesium: Green leafy vegetables, fish, meat, and dairy products.

Concentration/focus and nutrition

Exercise and nutrition may facilitate memory for older adults at high risk for developing

type 2 diabetes (Watson, Reger, Baker, & McNeely, 2006). Lifestyle modification creates a

compelling and accessible strategy to improve the quality of life, can prevent type 2 diabetes,

and can preserve cognitive functions (Watson et al., 2006). Endurance exercise and dietary fat

restriction improve metabolic parameters and verbal memory. These lifestyle interventions likely

influence memory through several central nervous system mechanisms related partly to

improved insulin regulation (Watson et al., 2006). Diet and exercise may also influence

neurotrophic factors and plasticity in brain regions directly relevant to memory (Watson et al.,

2006).

Anxiety and Nutrition

Diet modification can affect biological factors that influence the development of

depression (Jacka, Pasco, Mykletun, Williams, Hodge, O'Reilly & Berk, 2010). A study

examined the extent to which the high prevalence of mental disorders is related to habitual diet in

1,046 women ages 20-93 years randomly selected from the population (Jacka et al., 2010). After

adjustments for age, socioeconomic status, education, and health behaviors, a traditional dietary

pattern characterized by vegetables, fruit, meat, fish, and whole grains was associated with lower

odds of major depression or dysthymia and anxiety disorders (Jacka et al., 2010). A western

diet of processed or fried foods, refined grains, sugary products, and beer was associated with

more disorders (Jacka et al., 2010). Anxiety and other mental illness are related to decreased work productivity (Dewa, Thompson & Jacobs, 2011). Treatment for these disorders is significantly associated with productivity (Dewa et al., 2011).

Flu/illness and nutrition

Nutrition is a critical determinant of immune responses, and malnutrition is the most common cause of immunodeficiency worldwide (Chandra, 1997). Protein-energy malnutrition can impair cell-mediated immunity, phagocyte function, complement system, secretory immunoglobulin, A - a-antibody concentrations, and cytokine production (Chandra, 1997). The deficiency of single nutrients also results in altered immune responses, observed even when the deficit is relatively mild. Of the micronutrients, zinc; selenium; iron; copper; vitamins A, C; E, and B-6; and folic acid significantly influence immune responses. Overnutrition and obesity also reduce immunity (Chandra, 1997).

Oxidative stress plays a significant role in initiating and progressing many diseases, such as coronary heart disease, chronic heart failure, hyperlipidemia, hypertension, diabetes, cataracts, and cancer (Harasym & Oledzki, 2014). Oxidative stress arises when excessively formed free radicals exceed the capabilities of their neutralization by endogenous antioxidant systems, which can lead to illness (Harasym & Oledzki, 2014).

A decrease in plasma antioxidant capacity may indicate that the antioxidant defense system cannot keep up with the inactivation of oxygen free radicals and repair oxidative damage. Thus there is a risk of increased incidence of diseases (Harasym & Oledzki, 2014). Lower morbidity and illness correspond to a reasonable lifestyle and nutrition, which is dominant in consuming fruit and vegetables (Harasym & Oledzki, 2014).

Energy highs and lows

A meta-analysis of 88 studies examined the association between soft drink consumption and nutrition and health outcomes and found clear associations between soft drink intake with increased energy intake and body weight (Vartanian, Schwartz & Brownell, 2007). Soft drink intake was also associated with a lower milk intake, calcium, and other nutrients and an increased risk of several medical problems, such as diabetes (Vartanian et al., 2007). Lower soft drink consumption is shown to reduce health issues and balance energy.

Nearly all cases of mid-morning energy lows, afternoon lows, and general exhaustion at work are correlated to low blood sugar (Fitzpatrick, 2010). Coffee and a bagel in the morning give an intense but short energy boost. By mid-morning, energy will be depleted or there will be a desire to achieve another energy boost by consuming another carbohydrate energy boost. Instead of letting glucose bottom out around lunchtime, performance will be better by grazing throughout the day, as spikes and drops in blood sugar are both bad for productivity and harmful to the brain (Friedman, 2014). Smaller, more frequent meals maintain glucose at a more consistent level than relying on a midday feast (Friedman, 2014).

High G.I. foods, like white bread, white rice, and most cereals, are easily converted into glucose in the body (Fitzpatrick, 2010). Low G.I. foods like most vegetables, whole grains, meat, milk, nuts, and eggs convert much slower and keep the blood sugar more stable throughout the day (Fitzpatrick, 2010). Eating every 2-3 hours is optimal for maintaining stability and satiety. Eating a protein and complex carbohydrates snack between regular meals is best (Fitzpatrick, 2010). Cottage cheese, fruit, yogurt, tuna, hard-boiled eggs, protein bars, an apple with peanut butter, and trail mix are all healthy high-protein snacks (Fitzpatrick, 2010).

On activity, a wellness intervention at the workplace shows excellent results. Improvements in health behaviors, reductions in sickness absence, and improvements in job satisfaction and organizational commitment occurred following five years of a working

environment wellness intervention for employees (Blake, Zhou & Batt, 2013). At five years,

significantly more respondents actively traveled (by walking or cycling to work and for non-

work trips) and were active (Blake et al., 2014). Significantly more respondents met current

recommendations for physical activity at five years than at baseline (Blake et al., 2014). Fewer

employers reported a lack of time as a barrier to physically active following the intervention,

resulting in greater body energy levels (Blake et al., 2014). Significantly lower sickness absence,

greater job satisfaction, and more significant organizational commitment were reported at five

years than baseline (Blake et al., 2014).

Hydration. The minor signs of dehydration include lethargy, headaches, muscle pain,

and a general sense of confusion (Fitzpatrick, 2010). It is important to develop habits of regularly

drinking 32oz of water daily, which can lead to a more energized state (Fitzpatrick, 2010).

Providing water as the primary beverage at the workplace is an excellent way to promote

benefits and health (Feyerherm, Tibbits, Wang, Schram, & Balluff, 2014).

Sleep and productivity

Poor sleep decreases productivity and performance and increases mortality and morbidity

(Wells & Vaughn 2012). The National Sleep Foundation estimates poor sleep costs America

billions of dollars yearly and significantly compromises public safety and health (Wells &

Vaughn 2012). While individual sleep needs vary, the National Sleep Foundation notes that

adults typically need seven to nine hours of sleep each night (Barnes & Spreitzer, 2015). On any

given day, 29.9% of Americans report sleeping less than six hours the previous evening (Barnes

& Spreitzer, 2015). Sleep deprivation can lead to lower levels of effort, lower levels of

interpersonal helping behavior, and a higher prevalence of unethical and deviant behaviors

(Barnes & Spreitzer, 2015). A lack of sleep can also lead to lower trust and cooperation levels in negotiation processes (Barnes & Spreitzer, 2015).

Focusing on a task is more challenging when tired, and overall performance suffers. (Barnes & Spreitzer, 2015). Workplace injuries and greater injury severity are also linked to poor sleep patterns (Barnes & Spreitzer, 2015). Given how important it is for companies to control healthcare costs, it is also important to reiterate the link between healthy sleep patterns and health (Barnes & Spreitzer, 2015). Those suffering from chronic sleep loss are likelier to get sick, feel depressed, and be obese (Barnes & Spreitzer, 2015).

Making time for nutrition at work

It is important to make eating decisions before hunger occurs. If the decision is to go out for lunch, it is best to choose where in the morning versus right at lunchtime (Friedman, 2014). If ordering in, it is best to decide after a mid-morning snack, as studies show people are much better at resisting salt, calories, and fat in the future than in the present (Friedman, 2014).

Some ideas are placing a container of almonds and a selection of protein bars close by, near the line of vision, and bringing a bag of fruit to the office on Mondays to be available throughout the week (Friedman, 2014). Research indicates that eating fruits and vegetables throughout the day is not merely good for the body but also for the mind (Friedman, 2014).

Fruit and vegetable consumption predict greater well-being, curiosity, and creativity between and within-person levels (Conner, Brookie, Richardson, & Polak, 2015). Young adults who eat more F.V. reported higher average eudaemonic well-being, more intense feelings of curiosity, and greater creativity than those who ate less F.V. (Conner et al., 2015). Growing evidence shows that a diet rich in fruits and vegetables F.V. is related to greater happiness, life satisfaction, and positive affect (Conner et al., 2015).

Making choices and emotional eating

Emotional eating is when a person attempts to manage their mood with food, such as consuming large quantities of unhealthy foods in response to feelings instead of hunger (Weichert, 2015). When people cope with stress through compulsive and emotional eating, they create additional stress through what they practice (Weichert, 2015).

The non-diet approach to treating emotional eating helps individuals return to a normal relationship with food (Weichert, 2015). The concept of attuned eating addresses the questions of when and what to eat. Researchers, dieticians, and therapists who developed these ideas agree that humans have natural, internal instincts that can reliably direct them in self-regulating their food choices (Weichert, 2015).

Individuals who display symptoms of emotional eating have lost their innate ability to self-regulate their hunger and satiation (Weichert, 2015). Regardless of whether this lack of attunement comes from years of dieting, the use of food for affect regulation, or both, the first step in the process of curing emotional eating is for individuals to relearn how to listen accurately to and trust their bodies' signals (Weichert, 2015).

There are steps to Achieving Attuned Eating. These steps are: (Weichert, 2015).

1. Learn to identify physical hunger.

2. Determine the difference between physical hunger and emotional food craving.

3. Learn proper food-matching techniques.

4. Recognize and respond to food justification.

5. Reduce the desire for emotionally craved foods by stocking up on them.

6. Learn to identify physical fullness

In a particular study, results showed that subjective arousal seems to be related to food intake and appears to be affected by environmental stimuli such as the eating location and the number of people present (Stroebele & de Castro, 2006). Only minor effects of some environmental stimuli on heart rate correlate with higher average heart rates in restaurants (Stroebele & de Castro, 2006). Changes in arousal are not affected by ambient influences, but the environment and the individual's emotional state appear to play a role in the individual's eating behavior (Stroebele & de Castro, 2006).

The concept known as sociotropy involves a preoccupation with pleasing others and maintaining social harmony (Exline, Zell, Bratslavsky, Hamilton & Swenson, 2012). Sociotropy consists in trying to match someone's eating and eating to make the person feel comfortable (Exline et al., 2012). Sociotropy also can lead to more interpersonal concern and distress in these situations, giving in to social pressure by eating more (Exline et al., 2012).

Supplementation

Even though supplements may not offer all the benefits that whole foods can provide, there are times when taking vitamins and minerals in pill form may be appropriate (Vitamins and Minerals, 2009). An example is if the recommended servings of fruits, vegetables, and other healthy foods are lacking, there may be a benefit from a multivitamin that contains a variety of essential nutrients (Vitamins and Minerals, 2009). Multivitamins can also be helpful for a strict vegetarian or if someone eats a limited diet because of food allergies or intolerances or has a disease or condition that doesn't allow for digestion or absorption of nutrients properly (Vitamins and Minerals, 2009).

Older age and habits such as smoking and excessive alcohol consumption also can make it challenging to get all nutrients from food (Vitamins and Minerals, 2009). For pregnant women or those trying to become pregnant, nutrients like calcium, folic acid, and iron are needed to

protect the developing baby's health and health (Vitamins and Minerals, 2009). Supplementing

the diet with additional calcium and vitamin D is often considered crucial following menopause

to protect against osteoporosis and the risk of fractures (Vitamins and Minerals, 2009).

Getting too much of some nutrients from high-dose supplements can be dangerous,

especially with some fat-soluble vitamins. (Vitamins and Minerals, 2009). Some water-soluble

vitamins may also be toxic in large amounts (Vitamins and Minerals, 2009).

The 2005 Dietary Guidelines for Americans advises that nutrient needs should be met

primarily through consuming foods, with supplementation suggested for specific sensitive

populations (Fortify, 2015). These guidelines provide science-based advice to promote health

and reduce chronic disease risk through diet and physical activity. They form the basis for

federal food, nutrition education, and information programs (Fortify, 2015). Supplements may

be helpful when they fill a nutrient gap that cannot be achieved by food intake, but they are not a

substitute for a healthful diet (Fortify, 2015).

The 2010 Dietary Guidelines for Americans issued by the U.S. Department of

Agriculture and the U.S. Department of Health and Human Services recommend that nutrient

needs be met primarily through consuming foods. Also (Ward, 2014). In some instances,

fortified foods and dietary supplements may help provide one or more nutrients that otherwise

might be consumed in less than recommended amounts (Ward, 2014).

Multivitamin and mineral or M.V.M. supplements are not substitutes for a balanced diet,

and nutritional inadequacies are less common among those who take M.V.M. supplements.

M.V.M. supplement users may eat more nutrient-rich foods and have healthier lifestyles overall

(Ward, 2014).

It is not always possible for most people to choose foods containing the recommended amounts of all essential micronutrients, and chronic, relatively minor nutrient shortfalls can cause health problems (Ward, 2014). While M.V.M. supplements cannot replace eating adequate amounts of a variety of foods, they may be particularly beneficial to people with poor nutrition for some reasons, including inadequate intake of foods from all the food groups, advanced age, and chronic illness (Ward, 2014).

Overall, it is better to obtain nutrients from food containing various healthful ingredients. Fruits and vegetables contain many biologically active ingredients that may help to prevent diseases such as cancer in ways that vitamins and minerals alone do not (Ward, 2014).

Discussion

The effects that change in diet and lifestyle have on productivity are shown to have a convincing positive influence on overall well-being. Stress is a significant contributor, and the sudden release of stress on the body makes the body susceptible to illness. An example is when an individual gets sick on vacation after a sudden release of stress from the workplace. Regarding nutrition and productivity, it is beneficial to understand that choices and decisions create more substantial health and vitality.

The above concerns about diet and lifestyle are crucial regarding employee wellness and productivity. Much advice is available for new foods and lifestyle choices. These come from media such as television, social media, newspapers, and magazine ads. Most important is a choice and decision-making based on advice from reputable sources that come from professionals.

The concerns about these issues are what are the determinants of food and lifestyle choices and why people make the decisions they do. There are many factors, such as hunger, income, skills such as cooking, culture, mood, stress, education and attitudes, and beliefs about food (The determinants, 2016). The results of working with individuals who make choices and decisions to better their health are phenomenal. Many positive things can occur, such as raising morale, improving energy and productivity at work; promoting health and preventing disease, inspiring employees to take control of their health and well-being; creating a healthy organizational culture, and following the national benchmarks for a healthy workplace (The determinants, 2016).

Some studies factor in obesity at work and how to deal with it. Because obesity and illness are related to absence and reduced productivity, companies look for ways to reduce costs and positively contribute to the bottom line (Crash Course, 2006). In one example, a company can educate employees on body weight and its health effects. In another example, employers are advised to find out how many employees are overweight and get them to enter a program to get better (Crash Course, 2006).   This idea can raise issues of lowered self-esteem, making employees feel targeted and perhaps even considered being singled out.

Each point in the research section relates to mind/body/spirit. In Holistic nutrition, these work together to create homeostasis as a whole. The mind includes the senses, environment scanning, and emotions. Stress is one of the leading causes of lower productivity in the workplace, and there are ways the body can combat it effectively via proper nutrition and lifestyle. Lack of Concentration and focus is also known as brain fog.

A nutritional protocol alone won't cure anxiety, but overall lifestyle changes that include watching food consumption may help, as discussed with choice and decision-making. The body is the entire quantifiable or tangible structure of a human being. It is the visible portion of an individual. Sleep, illness, and energy levels fall under the mind category. Lack of sleep is a common issue for many people. Things such as waking up in the middle of the night to urinate, night sweats, and problems with sleep may arise from lifestyle and nutrition. On the immune system, there are ways to help the body become more resilient to the flu virus via proper lifestyle and diet to help promote stronger immunity. Energy highs and lows throughout the day. There are various ways that adequate nutrition and intake of different foods can prevent energy ups and downs throughout the day.

Spirit/Open Awareness relates to the consciousness or personality of an individual. Fitting in nutrition and emotional eating works with being conscious of what a person is doing and making the choices to be healthy. As discussed, many protocols help support proper nutrition throughout the day. An example is consuming fruits and vegetables versus sugary snacks. During meetings, adequate nutrition and bringing in whole foods versus processed foods are part of deciding on intelligent choices. Many of the points are tied to emotional eating, which can be attributed to mindless eating, eating while bored, and life experiences.

Conclusion and recommendations

As organizations become more involved with today's demands, it is vital to understand how nutritional health and lifestyle benefits employee productivity and effectiveness. Wellness is essential, the return on investment for productive employees, and health is a condition of mental and physical well-being relating to productivity and performance.

Nutrition plays a critical role since nutrient deficiencies lead to compromised mental and physical states, resulting in bad health and poor work performance. Proper nutrition and lifestyle are also central to preventing and controlling diseases. Health and lifestyle can impact how food contributes to positive social, mental, and physical well-being (Aldinger & Jones, 1998). Time, attention, and support in the household and the workplace are necessary to meet people's food and health needs (Aldinger & Jones, 1998). Social ties are also validated and maintained by the exchange of food, for offering food is associated with providing love, affection, and friendship (Aldinger & Jones, 1998).

For higher productivity, employers and employees must collaborate and discuss points in this paper. It is essential to understand the various issues affecting people, raise these concerns, and get to the cause. As in the research, stress is one of the highest issues on productivity. It can lead to many illnesses and fatigue. Coping mechanisms that can be nutritional is critical for dealing with stress and thus can alleviate productivity concerns.

Many forms of media, such as the internet, television, and magazines, deliver nutrition information that people may interpret differently. Mass media is ubiquitous and powerful and can increase body dissatisfaction among people (Derenne & Beresin, 2006). Mass media can encourage new behaviors by being used for health promotion and nutrition interventions. Mass media can include signboards, posters, calendars, magazines, booklets, leaflets, and audiovisual materials (Aldinger & Jones, 1998).

A support system to reinforce education and promote health can be very beneficial. People will likely embrace healthy lifestyle choices if constant information and support are collaborative. Each person has their health journey, and it is crucial to understand each person's journey to help with success. Promoting healthy eating and moderate physical activity can increase self-esteem, and funding for high-quality and visible advertising campaigns promoting healthy lifestyles can raise awareness (Derenne & Beresin, 2006).

Eating healthy doesn't necessarily mean good health. Although a healthy diet is part of a positive lifestyle, it also includes incorporating exercise and avoiding substance abuse. (Aldinger

& Jones, 1998). The importance of holistically addressing lifestyle consists of the body, nutrition, and active living. It is beneficial to discuss fitness and nutrition together.

Evaluation can inform and strengthen workplace health programs (Aldinger & Jones, 1998). Assessments provide information about how well the program works and its intended effect. Evaluation is a critical program element that can be implemented from the start and remains ongoing. Evaluation helps with things such as giving updated information to policymakers, planners, and participants about the program's effects, providing feedback to those involved, and thus leading to improvements and adjustments during the process.

A nutritional program is a critical intervention to help promote healthy eating. Nutritional seminars are beneficial at the workplace, discussing healthy snacks and dietary ideas. Employees and their families can be involved with this program by taking part in workshops to gain knowledge about nutrition and skills such as cooking. Nutritionists from the community can also be at the workplace for lunch and learns, offer nutritional assessments to staff, and even do supermarket tours with groups. Community services dedicated to these needs in the community can provide guidance and encouragement on nutrition and health protocols. Advocating for health is vital as it takes knowledge and skill that people can learn from, resulting in sustainable, healthy lifestyle routines.

Each section of wellness ties in very well together. Dealing with work travel and vacation ties in with stress and anxiety, and the associated protocols are also closely related. Also, it ties into energy highs and lows throughout the day and fits in proper nutrition and supplementation at work and home. For the entire protocol to be helpful for each wellness section, emotional eating needs to be conquered. Choice and decisions on food consumption are closely related to emotional eating. With help and perseverance, this can be achieved.

With active support, education and ongoing evaluation for the long term, more substantial productivity and strengthened employee engagement can happen. The impact on productivity can be tremendous. This paper notes that low production costs can be up to 576 billion dollars USD. With rising healthcare costs, it is crucial to deal with the causes. Not only will this positively affect workplace productivity, but it will also create a healthier society for future generations. Therefore, the wellness sections in this document relate to lowering the costs of an unproductive workplace and thus saving billions of dollars for employees, employers, and the economy.

References

5 Factors that affect your employee's productivity. (2016). Retrieved from

https://www.nbrii.com/employee-survey-white-papers/5-factors-that-affect-your-

employees-productivity/

A.C.S.; wellness survey: Businesses look to wellness programs to improve productivity and

lower absenteeism. (2009). *Fitness & Wellness Business Week,* 9. Retrieved from

http://search.proquest.com/docview/214044747?accountid=158302

Aldinger, C., & Jones, J. (1998). Healthy Nutrition: An Essential Element of a Health-

Promoting School. Retrieved from http://www.who.int/hpr

Alpert, J.E., Mischoulon, D., Nierenberg, A.A., Fava, Maurizio. (2000). nutrition and

depression: focus on folate. *Nutrition,* 16(7-8), 544-546. DOI:

http://dx.doi.org/10.1016/S0899-9007(00)00327-0. Retrieved from

http://www.nutritionjrnl.com/article/S0899-9007 (00)00327-0/fulltext

Barnes, C. M., & Spreitzer, G. (2015). Why sleep is a strategic resource. *M.I.T. Sloan

Management Review, 56*(2), 19-21. Retrieved from

http://search.proquest.com/docview/1650910792?accountid=158302

Beling, J. (2007). The anti-aging fitness prescription: A day-by-day nutrition and workout plan to

age-proof your body and mind. *Physical Therapy, 87*(9), 1255-1256. Retrieved from

http://search.proquest.com/docview/223124899?accountid=158302

Blake, H., Zhou, D., & Batt, M. E. (2013). Five-year workplace wellness intervention in the

N.H.S. *Perspectives in Public Health, 133*(5), 262-71. Retrieved from

http://search.proquest.com/docview/1438039939?accountid=158302

Bowman, J. (2014). 7 Health Claims About Astaxanthin. Retrieved from

http://www.healthline.com/health-slideshow/health-claims-astaxanthin#1

Brown, M. (2014). Flying high. Retrieved from

http://melaniebrownnutrition.com/2014/02/03/flying-high/

Ceniceros, R. (2012). Tracking True Cost of Lost Productivity Remains a Challenge. Retrieved

from http://www.workforce.com/articles/tracking-true-cost-of-lost-productivity-remains-

a-challenge

Chandra, R.K. (1997). Nutrition and the immune system: an introduction. *The American Journal

of Clinical Nutrition,* 66(2), 4605-4635. Retrieved from

http://ajcn.nutrition.org/content/66/2/460S.short

Conner, T.S., Brookie, K.L., Richardson, A.C., & Polak, M.A. (2015). On carrots and curiosity:

eating fruit and vegetables is associated with greater flourishing in daily life. *British

Journal of Health Pychology,* 20(2), 413-427. doi: 10.1111/bjhp.12113. Retrieved from

http://www.ncbi.nlm.nih.gov/pubmed/25080035

Crash course in … Tackling Obesity at Work (2006). Retrieved from

https://www.questia.com/magazine/1G1-155473967/crash-course-in-tackling-obesity-at-

work

Derenne, J. L., & Beresin, E. V. (2006). Body image, media, and eating disorders. *Academic

Psychiatry, 30*(3), 257-61. Retrieved from

http://search.proquest.com/docview/196508089?accountid=158302

Dewa, C., Thompson, A. H., & Jacobs, P. (2011). The association of treatment of depressive

episodes and work productivity. *Canadian Journal of Psychiatry, 56*(12), 743-50.

Retrieved from http://search.proquest.com/docview/916423797?accountid=158302

Ehrlich, S.D. (2011). Nutrition. Retrieved from

http://umm.edu/health/medical/altmed/treatment/nutrition

Exline, J. J., Zell, A. L., Bratslavsky, E., Hamilton, M., & Swenson, A. (2012). People-pleasing

through eating: Sociotropy predicts greater eating in response to perceived social

pressure. *Journal of Social and Clinical Psychology, 31*(2), 169-193.

doi:http://dx.doi.org/101521jscp2012312169

Feyerherm, L., Tibbits, M., Wang, H., Schram, S., & Balluff, M. (2014). Partners for a healthy

city: Implementing policies and environmental changes within organizations to promote

health. *American Journal of Public Health, 104*(7), 1165-8. Retrieved from

http://search.proquest.com/docview/1545530750?accountid=158302

Fitzpatrick, J. (2010). Eat your way to a high energy work day. Retrieved from

http://lifehacker.com/5664322/eat-your-way-to-a-high-energy-workday

Fortify Your Knowledge About Vitamins. (2015). Retrieved from

http://www.fda.gov/ForConsumers/ConsumerUpdates/ucm118079.htm

Friedman, R. (2014). What You Eat Affects Your Productivity. Retrieved from

https://hbr.org/2014/10/what-you-eat-affects-your-productivity

Gorman, K. (2009). Slender advantage. *The Safety & Health Practitioner, 27*(6), 51-52, 4.

Retrieved from http://search.proquest.com/docview/201045138?accountid=158302

Hall, K.D., Heymsfield, S.B., Kemnitz, J.W., Klein, S., Schoeller, D.A. & Speakman, J.R.

(2002). Energy balance and its components: implications for body weight regulation. *The

American Journal of Clinical Nutrition,* 95(4), 989-994. Retrieved from

http://ajcn.nutrition.org/content/95/4/989

Harasym, J., & Oledzki, R. (2014). Effect of fruit and vegetable antioxidants on total antioxidant

capacity of blood plasma. *Nutrition, 30*(5), 511-7.

doi:http://dx.doi.org/10.1016/j.nut.2013.08.019

Harrison, A., Sullivan, S., Tchanturia, K., & Treasure, J. (2010). Emotional functioning in eating

disorders: Attentional bias, emotion recognition and emotion regulation. *Psychological

Medicine, 40*(11), 1887-97. doi:http://dx.doi.org/10.1017/S0033291710000036

Holwegner, A. (2013). Encouraging Healthy Habits Can Save Employers Time and Money.

Retrieved from http://www.healthstandnutrition.com/encouraging-healthy-habits/

Jacka, F. N., Pasco, J. A., Mykletun, A.,Williams, L. J., Hodge, A. M.,O'Reilly, S. L.,... Berk,

M. (2010). Association of western and traditional diets with depression and anxiety in

women. *The American Journal of Psychiatry, 167*(3), 305-11. Retrieved from

http://search.proquest.com/docview/220493250?accountid=158302

Jefferies, S.C. (1999). Qualitative Research. Retrieved from

http://www.cwu.edu/~jefferis/PEHL557/pehl557_qualit.html

Jones, C. (2014). Why you always get ill on holiday and how to fight it when you do. Retrieved

from http://www.mirror.co.uk/news/uk-news/you-always-ill-holiday-how-3721672

Kroner, Z. (2011). Maximizing Health and Well-Being While Flying. Retrieved from

http://acam.typepad.com/blog/2011/03/maximizing-health-and-well-being-while-

flying.html

Leonard, B. (2013). How does diet impact health? Retrieved from

http://www.takingcharge.csh.umn.edu/enhance-your-wellbeing/health/diet-nutrition/how-

does-diet-impact-health

Lowden, A., Moreno, C., Holmbäck, U., Lennernäs, M., & Tucker, P., (2010). Eating and shift

work - effects on habits, metabolism, and performance. *Scandinavian Journal of Work,*

Environment & Health, 36(2), 150-62. Retrieved from

http://search.proquest.com/docview/201391973?accountid=158302

Mahr, D., Kalogeras, N., & Odekerken-Schröder, G. (2013). A service science approach for

improving healthy food experiences. *Journal of Service Management, 24*(4), 435-471.

doi:http://dx.doi.org/10.1108/JOSM-04-2013-0089

Lassen, A. D., Thorsen, A. V., Sommer, H. M., Fagt, S., Trolle, E., Biltoft-Jensen, A., & Tetens,

I. (2011). Improving the diet of employees at blue-collar worksites: Results from the

'food at work' intervention study. *Public Health Nutrition, 14*(6), 965-74.

doi:http://dx.doi.org/10.1017/S1368980010003447

Linton, S. (2013). Creating a workplace wellness program. Retrieved from

https://charityvillage.com/Content.aspx?topic=Creating_a_workplace_wellness_program

#.VrOR-4-cHIU

Nutrition and Productivity. (2013). Retrieved from http://www.healthline.com/health/nutrition-

and-productivity#2

Nutrition and Stress. (n.d.). Retrieved from https://uwaterloo.ca/health-services/nutrition-

services/nutrition-resources/nutrition-and-stress

Nutrition and Wellness. (2016). Retrieved from http://www.uwhealth.org/nutrition-

wellness/nutrition-and-wellness/11492

Popkin, B. M. (2005). Using research on the obesity pandemic as a guide to a unified vision of

nutrition. *Public Health Nutrition, 8*(6), 724-9.

doi:http://dx.doi.org/10.1079/PHN2005776

Pringle, A., Harmer, C. J., & Cooper, M. J. (2010). Investigating vulnerability to eating

disorders: Biases in emotional processing. *Psychological Medicine, 40*(4), 645-55.

doi:http://dx.doi.org/10.1017/S0033291709990778

Ranking - Best Multivitamins in 2016. (2016). Retrieved from

https://www.multivitaminguide.org/

Raulio, S., Roos, E., & Prättälä, R. (2010). School and workplace meals promote healthy food

habits. *Public Health Nutrition, 13*(6), 987-92.

doi:http://dx.doi.org/10.1017/S1368980010001199

Ritchie, J. (n.d.). Retrieved from http://www.stress.org.uk/files/combat-nutritional-stress.pdf

Seccombe, I. (1995). Sickness absence and health at work in the N.H.S. *Health Manpower

Management, 21*(5), 6. Retrieved from

http://search.proquest.com/docview/206621291?accountid=158302

Stress effects on the body. (2016). Retrieved from http://www.apa.org/helpcenter/stress-

body.aspx

Stroebele, N., & de Castro, J. M. (2006). Influence of physiological and subjective arousal on

food intake in humans. *Nutrition, 22*(10), 996-1004.

doi:http://dx.doi.org/10.1016/j.nut.2006.07.003

Study: Poor employee health habits result in lost productivity. (2012). *Health & Beauty Close -

Up,* Retrieved from http://search.proquest.com/docview/1032701357?accountid=158302

The Determinants of Food Choice. (2016). Retrieved from

http://www.eufic.org/article/en/expid/review-food-choice/

Vartanian, L. R., Schwartz, M. B., & Brownell, K. D. (2007). Effects of soft drink consumption

on nutrition and health: A systematic review and meta-analysis. *American Journal of*

Public Health, 97(4), 667-75. Retrieved from

http://search.proquest.com/docview/215084540?accountid=158302

Vitamins and Minerals What You Should Know About Essential Nutrients. (2009). Retrieved

from http://www.mayoclinic.org/documents/mc5129-0709-sp-rpt-pdf/doc-20079085

Ward, E. (2014). Addressing nutritional gaps with multivitamin and mineral supplements.

Nutrition Journal, 13(72), 1-10. Retrieved from

https://nutritionj.biomedcentral.com/articles/10.1186/1475-2891-13-72

Watson, G. S., Reger, M. A., Baker, L. D., McNeely, M. J., & al, e. (2006). Effects of exercise

and nutrition on memory in Japanese Americans with impaired glucose tolerance.

Diabetes Care, 29(1), 135-6. Retrieved from

http://search.proquest.com/docview/223042668?accountid=158302

Wells, M. E., & Vaughn, B. V., (2012). Poor sleep challenging the health of a nation. *The*

Neurodiagnostic Journal, 52(3), 233-49. Retrieved from

http://search.proquest.com/docview/1152078017?accountid=158302

Weichert, K. (2015). Breaking Emotional Eating Barriers, lecture notes. Retrieved from

file:///C:/Users/John/Documents/nutritional%20info/335_-

_Breaking_Emotional_Eating_Barriers_with_SGT_KEN.pdf

Why Wellness Matters: The Real Cost to Employers of Unhealthy Employee Behaviors. (2013).

Retrieved from

http://www.advisorperspectives.com/commentaries/mp_062013.php?channel=Retirement

Young, S. N. (2002). Clinical nutrition: 3. the fuzzy boundary between nutrition and

psychopharmacology. *Canadian Medical Association.Journal, 166*(2), 205-9. Retrieved

from http://search.proquest.com/docview/204820529?accountid=158302

YOUR KNOWLEDGE HAS VALUE

- We will publish your bachelor's and
  master's thesis, essays and papers

- Your own eBook and book -
  sold worldwide in all relevant shops

- Earn money with each sale

Upload your text at www.GRIN.com
and publish for free